AF822062

Science Word Gallery: CELLS

Mizzz Foster

Copyright © 2025 Melanie A. Foster
All rights reserved.

This book, or parts thereof may not
be reproduced in any form without
permission from the publisher; exceptions
are made for brief excerpts used
in published reviews.

Published by Fantastic Absurdities
Lake Elsinore, CA US
Email: MizzzFoster@gmail.com

ISBN: 978-1-7377913-4-8

Printed in the United States of America
Design by M. A. Foster
Cover Design by M. A. Foster
Illustrations © 2024 M. A. Foster

Welcome to *Science Gallery of Words: Cells*! This book is designed to help you remember and understand essential cell terminology in a fun and engaging way. It can be used in the classroom, at home, as a thoughtful gift for a science enthusiast, or simply for colorful enjoyment. You can even take the pages apart to create a colorful word wall. However, you choose to use *Science Gallery of Words: Cells*, my hope is that it inspires curiosity and deepens your love of science.

Cells: Cells are the basic structural and functional units of living organisms. There are two types of cells: prokaryotes and eukaryotes. Prokaryotes make up bacteria cells and are simple with no membrane bound organelles. Eukaryotes make up Protists, Animals, Plants, and Fungus. Eukaryotic cells are complex and filled with membrane bound organelles.

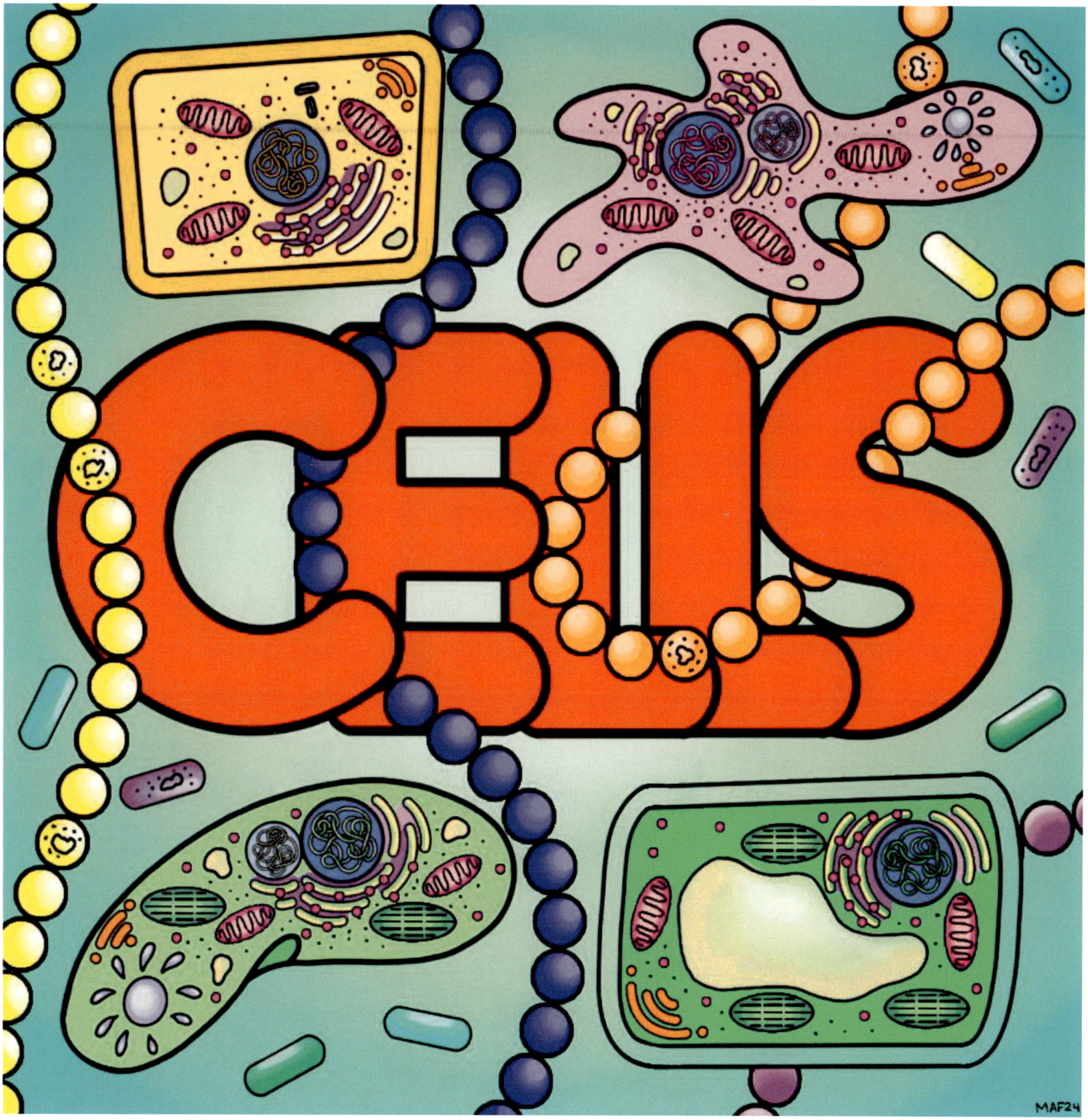
CELLS
MAF24

Prokaryote: Simple cells which do not contain a membrane bound nucleus and membrane bound organelles. Prokaryotes are bacteria cells and make up the Archaebacteria and Eubacteria kingdoms. “Pro” means before, and “Karyon” means kernel which symbolizes the nucleus. Prokaryotes are the first cells to evolve, the eukaryotes followed later.

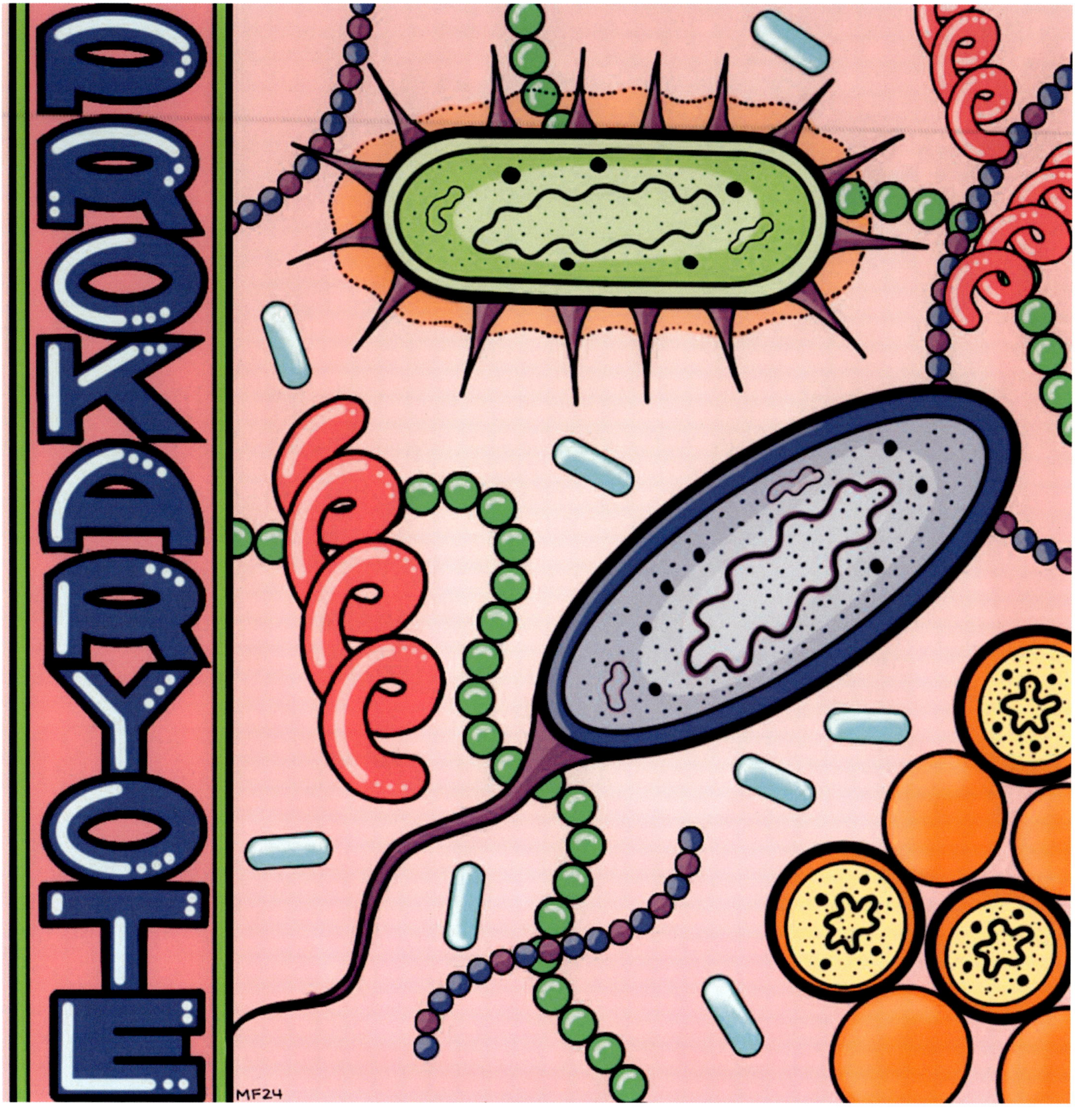
PROKARYOTE
MF24

Eukaryote: Complex cells which contain a membrane bound nucleus and membrane bound organelles. Eukaryotes are complex cells because they have a wide variety of organelles and can be classified into four kingdoms: Protist, Plant, Animal, and Fungus. "Eu" means true, and "Karyon" means kernel which is how the nucleus was described when the cells were viewed under a microscope.

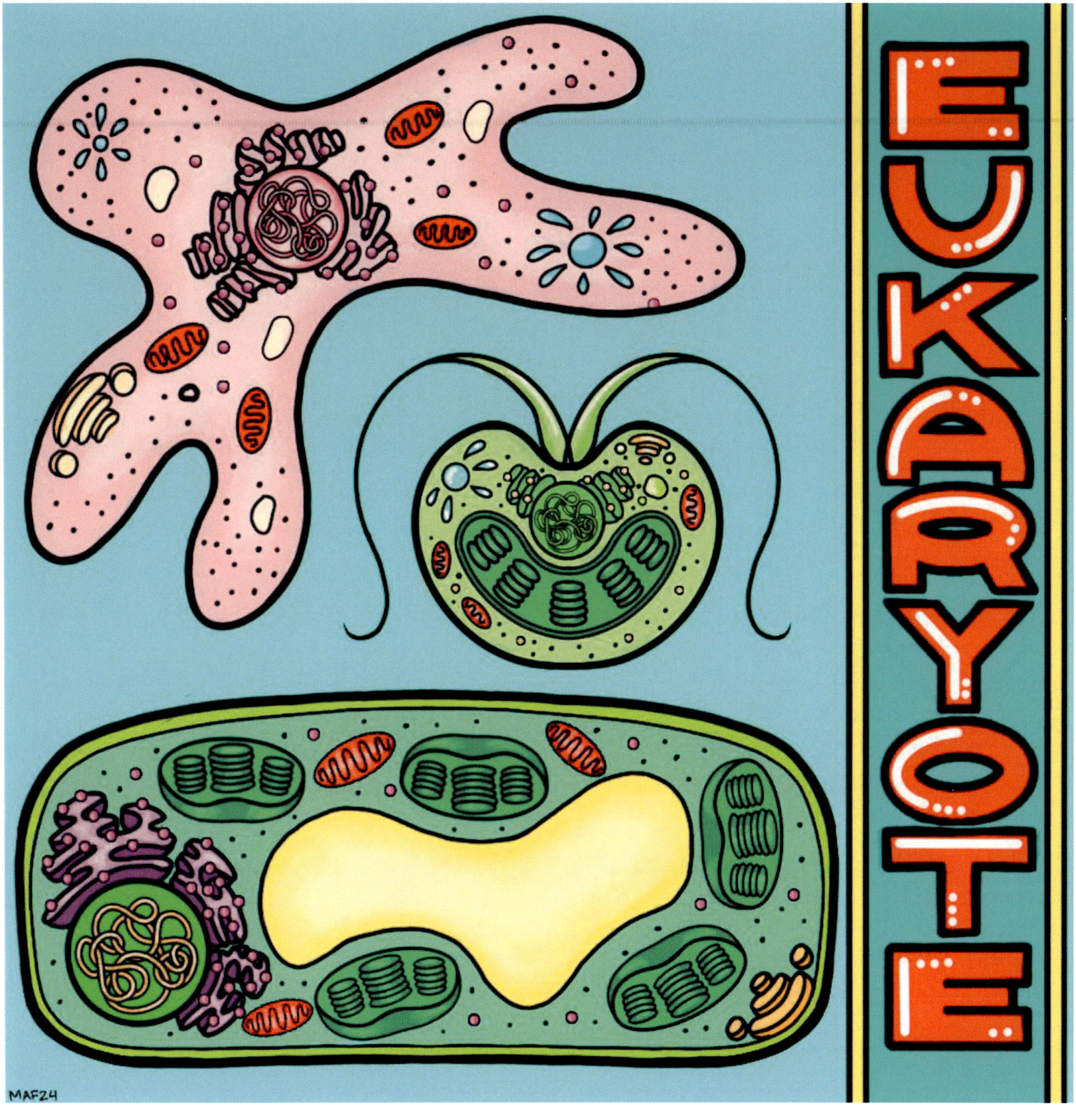
EUKARYOTE
MAF24

Cell membrane: Also called the plasma membrane. The cell membrane is a thin, semi-permeable barrier that surrounds the cytoplasm of all cells. It separates the interior of the cell from the external environment and regulates the passage of substances such as water, glucose, and ions in and out of the cell. The cell membrane is a crucial component of cells, playing essential roles in maintaining cell structure, communication, and homeostasis.

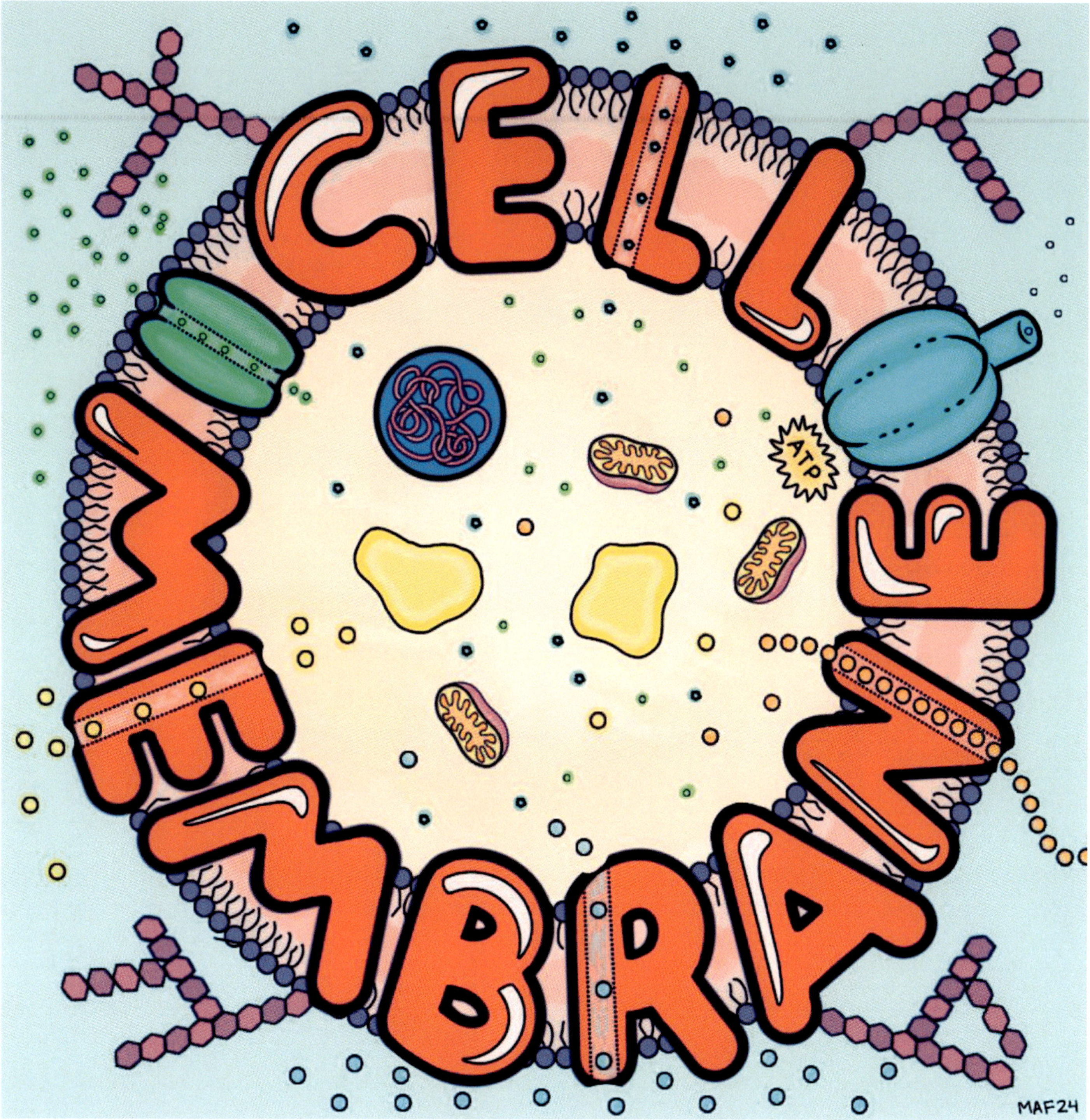
CELL
MEMBRANE
ATP
MAF24

Phospholipid: Biomolecule composed of two hydrophobic fatty acid chains and a hydrophilic head made up of glycerol and phosphate. Phospholipids create a semi-permeable bilayer that is the main component of cell membranes (plasma membranes).

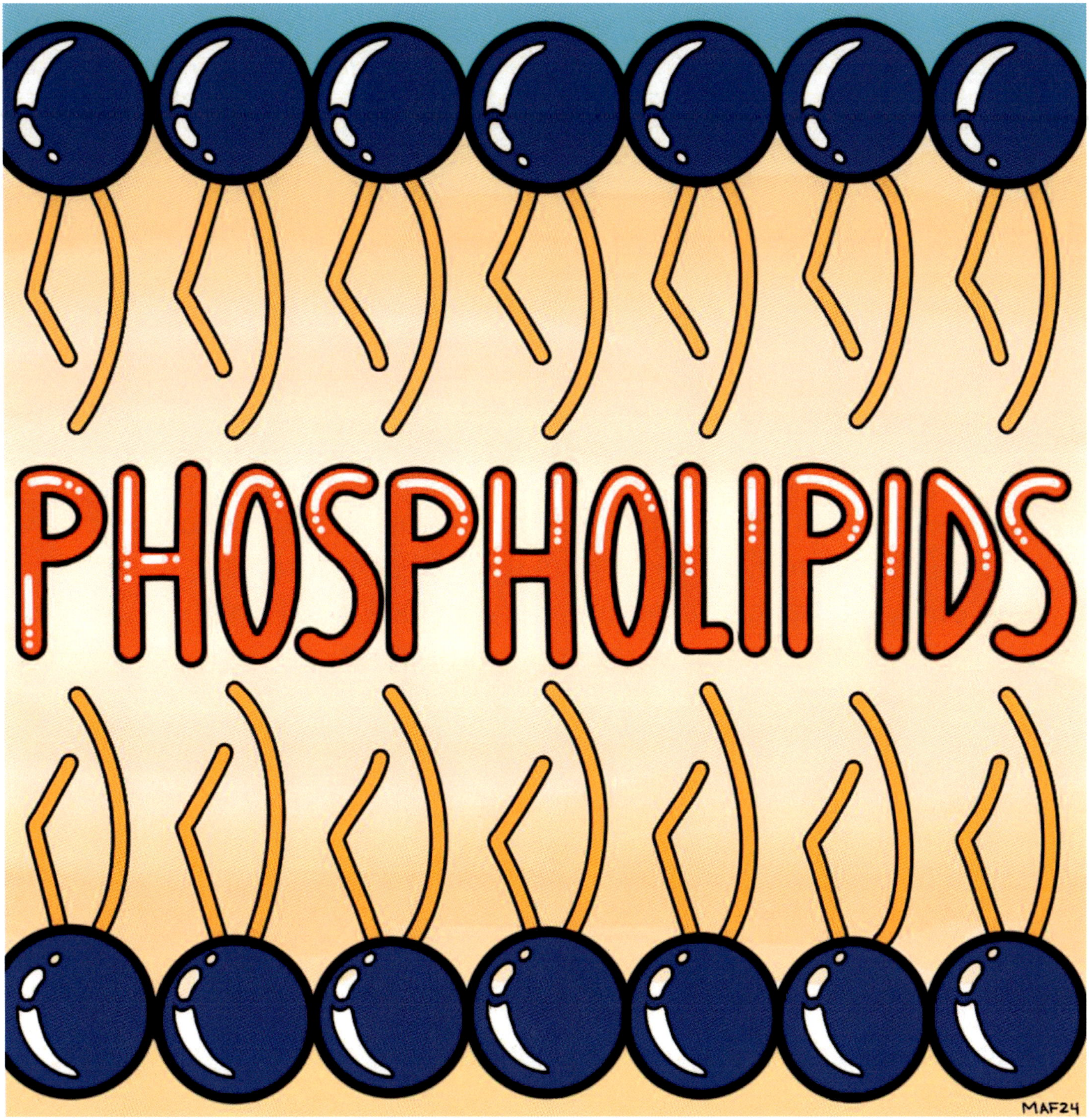
PHOSPHOLIPIDS
MAF24

Cytoplasm: The gel-like semi-fluid matrix composed of water, minerals, and various organic compounds such as proteins, lipids, carbohydrates, and nucleic acids. Cytoplasm supports the cell's structure through the cytoskeleton, serving as a medium for intracellular transport and communication.

CYTOPLASM
MF24

Cytoskeleton: A complex network of long thin proteins categorized as microtubules, intermediate filaments, or microfilaments found in eukaryotes. The cytoskeleton provides the cell with structural support, maintains the cell's shape, facilitates cellular movements and mediates intracellular transport.

CYTOSKELETON
MAF24

Microtubules & Microfilaments: Cytoskeleton filaments found in eukaryotic cells which play an integral part in maintaining the cell shape, helping with cell motility, and intracellular transport.

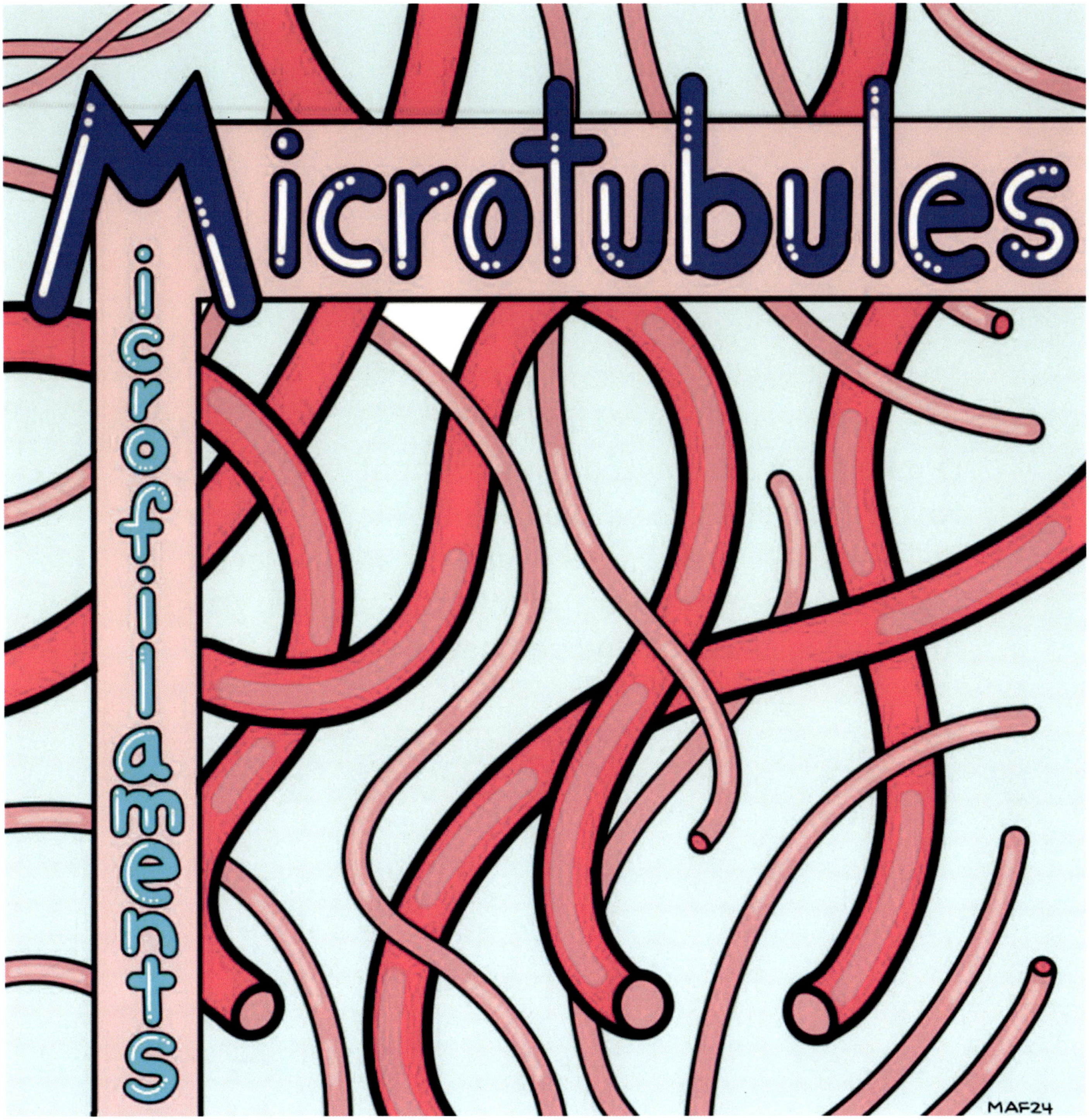
Microtubules
Microfilaments
MAF24

Organelle: A specialized compartment within cells which has a specific structure and function. Most organelles are membrane-bound except for ribosomes. Organelles perform specific activities to aid in growth, functioning, survival, and defense of cells.

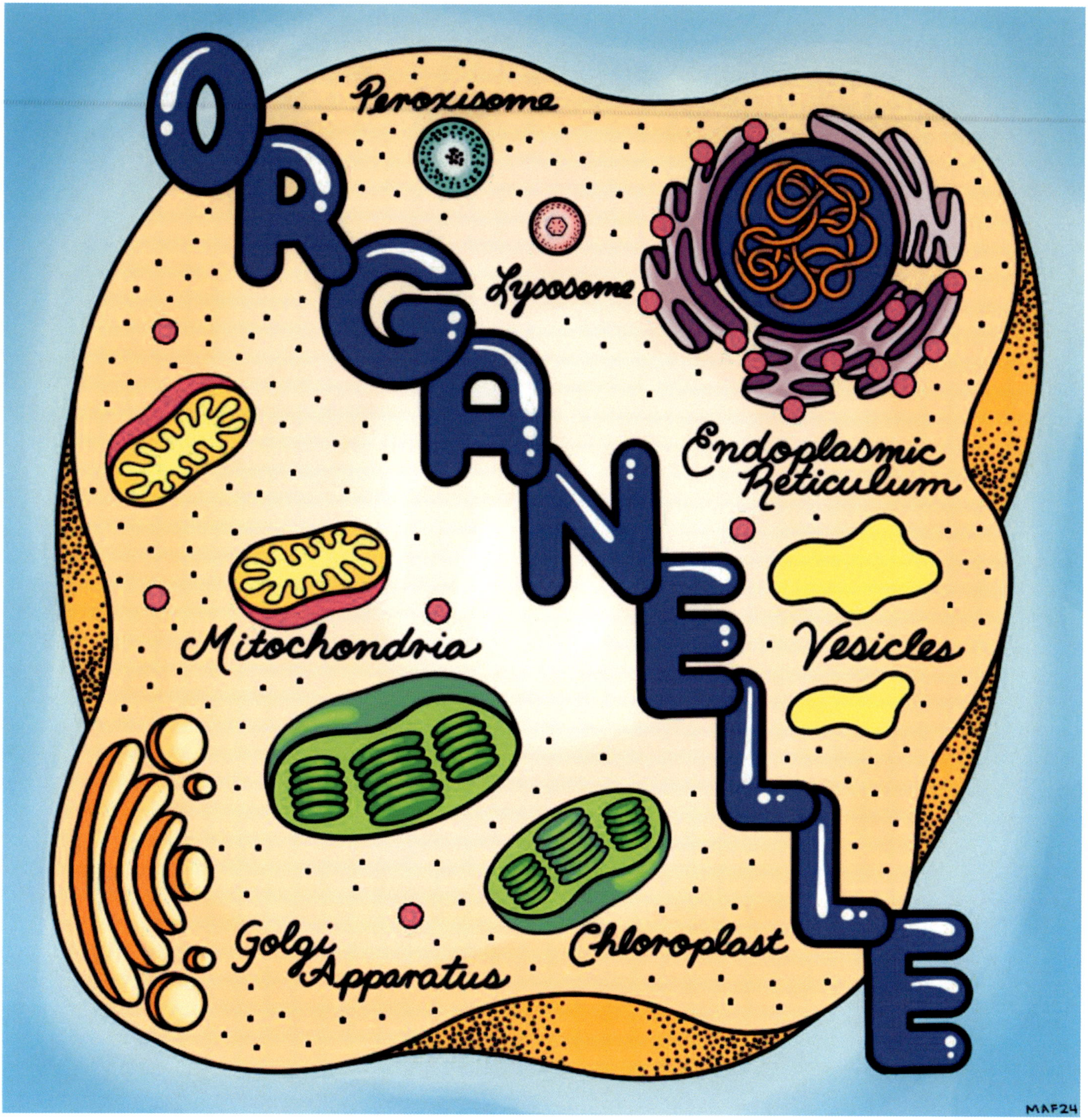
ORGANELLE
Peroxisome
Lysosome
Endoplasmic Reticulum
Mitochondria
Vesicles
Golgi Apparatus
Chloroplast
MAF24

Nucleus: Membrane-bound organelle found in eukaryotes which houses the cell's genetic information as chromosomes. The nucleus is the site for DNA synthesis and DNA transcription into mRNA. The nucleus is also where the nucleolus is housed. The nucleolus produces ribosomes.

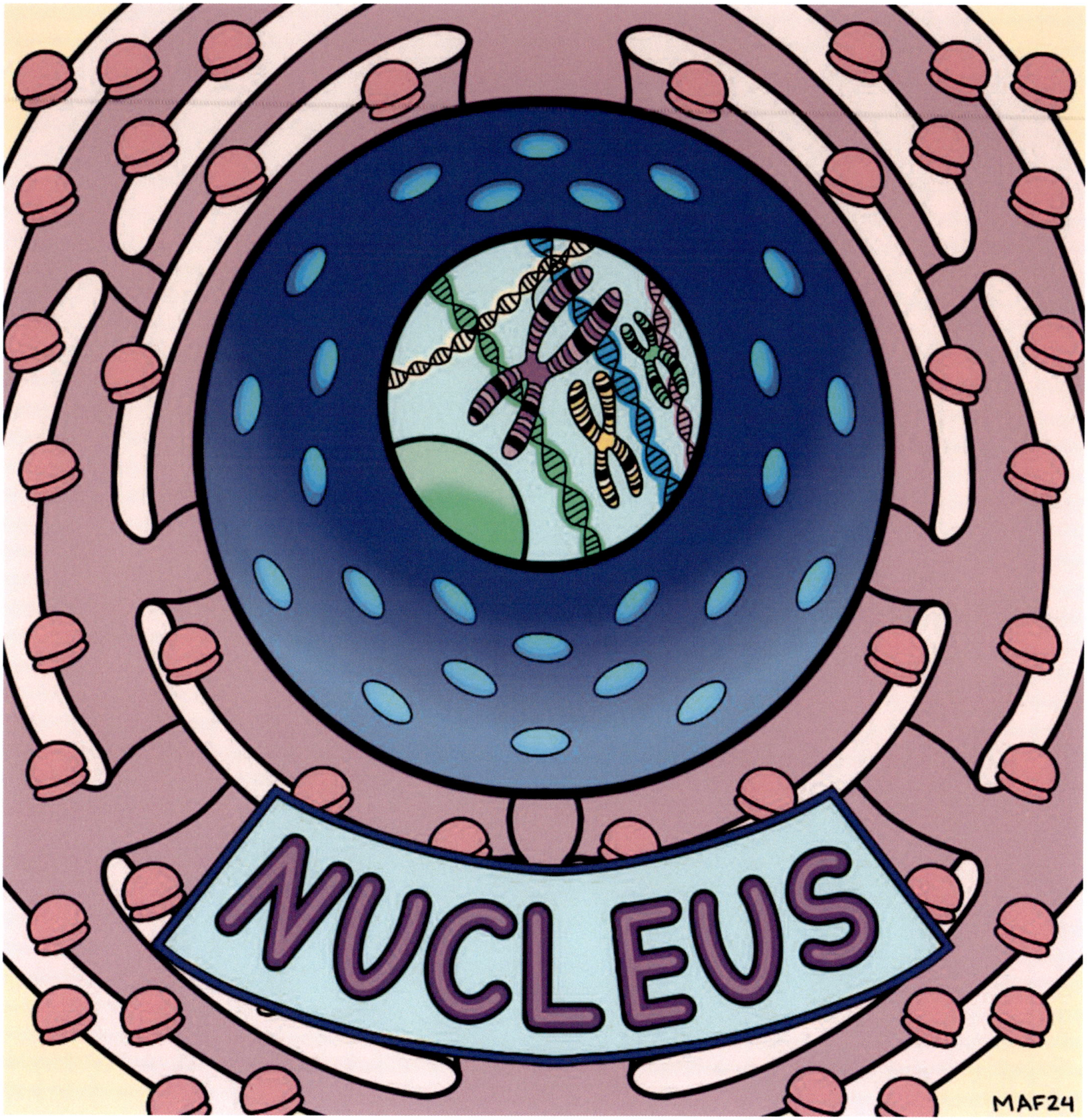
NUCLEUS
MAF24

Endoplasmic reticulum: A network of membrane-bound tubules and sacs that extend throughout the cytoplasm of eukaryotic cells. There are two types of ER, rough endoplasmic reticulum (RER) which has ribosomes attached and smooth endoplasmic reticulum (SER) which does not have ribosomes attached. RER is involved with the synthesis of phospholipids and processing of proteins. SER is involved with lipid metabolism and synthesis. SER also stores calcium ions.

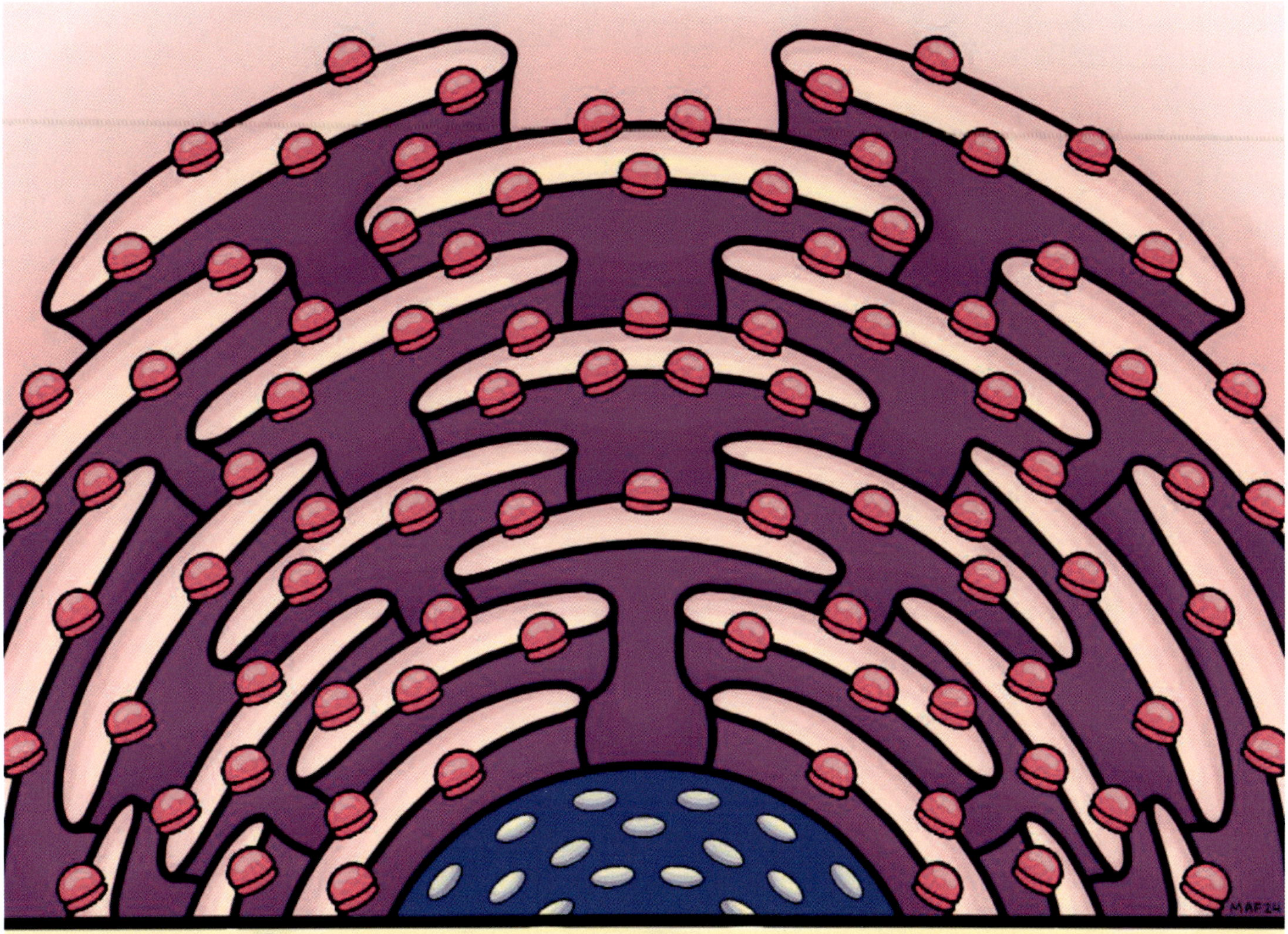

ENDOPLASMIC RETICULUM

Ribosome: Nonmembrane-bound organelle found in both prokaryotes and eukaryotes which synthesize protein.

Ribosome
large subunit
small subunit
MAF24

Golgi apparatus: Membrane-bound stack of cisternae found in eukaryotes which process, sort, and package proteins and lipids that are made within the cell. The Golgi apparatus transports lipids and proteins out of the cell through exocytosis.

GOLGI
APPARATUS
MAF24

Vacuole: Membrane-bound organelles which store water and other important nutrients and biomolecules in eukaryotic cells. The large central vacuole found in plant cells maintain turgor pressure by storing water which gives the plant support to stand upright and maintain healthy firm stems and leaves.

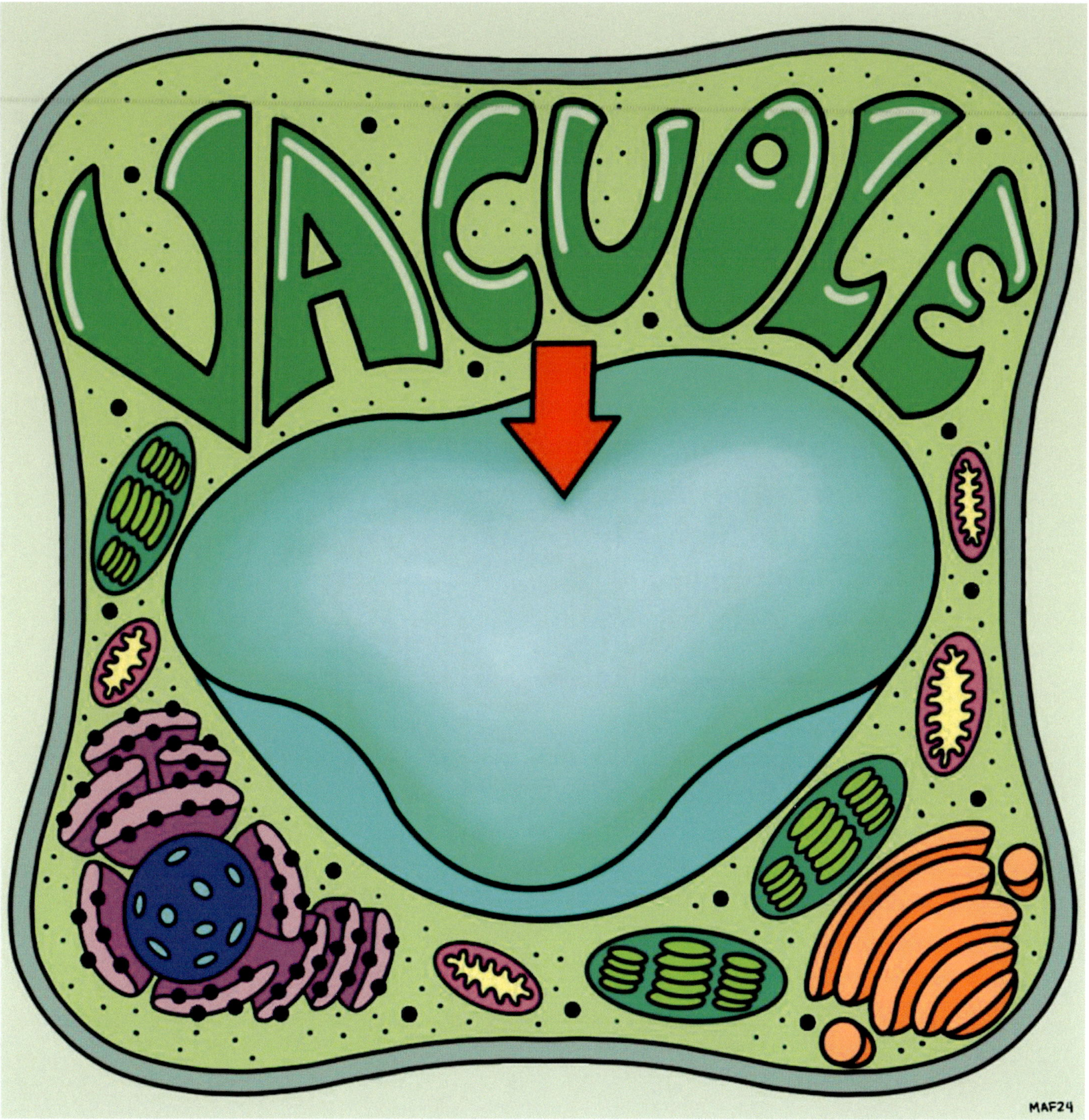
VACUOLE
MAF24

Cell wall: A rigid, protective layer that surrounds the cell membrane of many types of cells. The cell wall provides structural support, protection, and shape to the cell. Cell walls are primarily present in plant cells, fungal cells, bacterial cells, and some protist cells, but they are not found in animal cells.

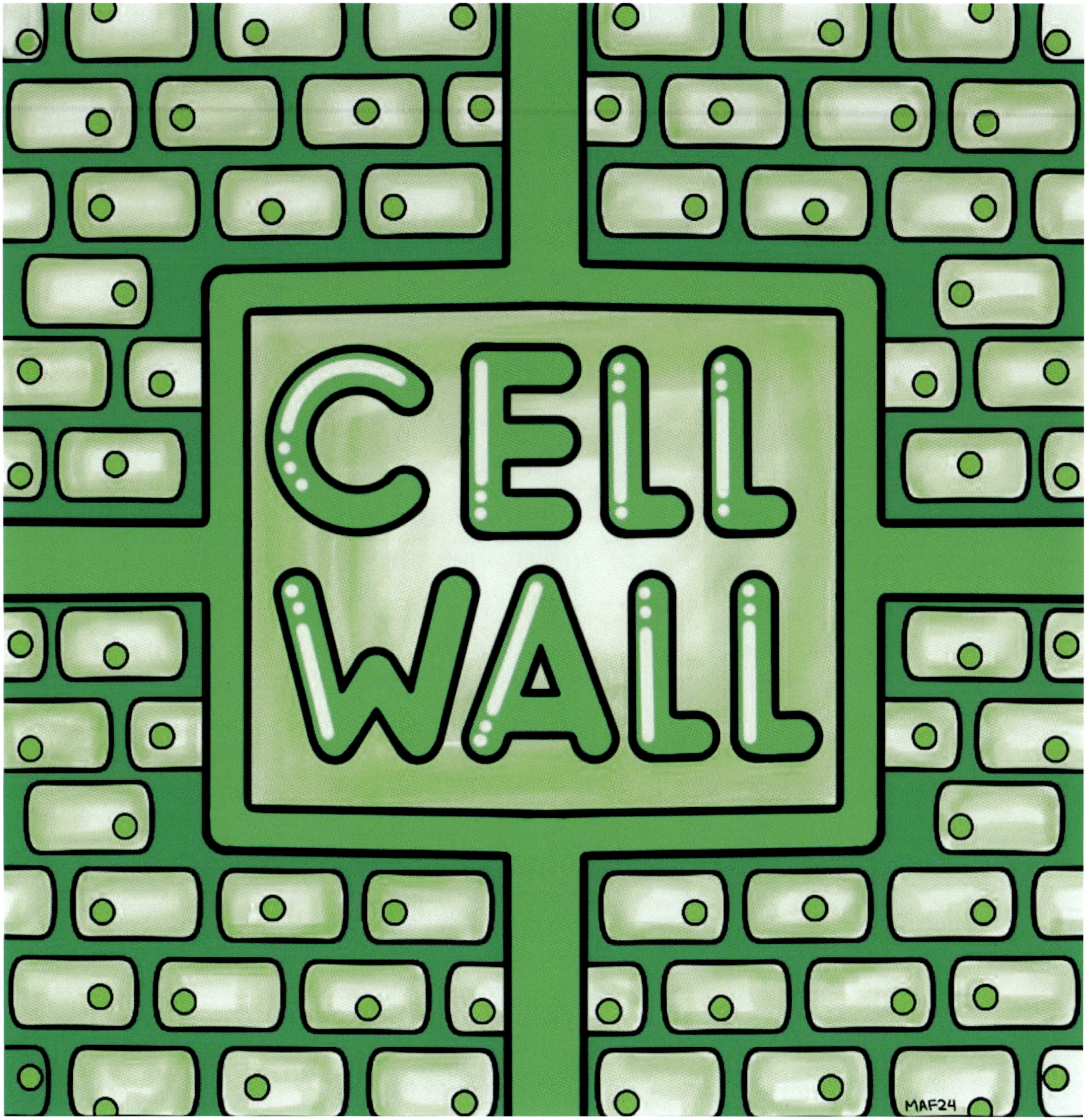
CELL
WALL
MAF24

Mitochondria: Membrane-bound organelle found in eukaryotes which is the site for ATP production during cellular respiration (aerobic respiration). ATP is the source of energy for cells, which is why mitochondria are often referred to as the powerhouse of the cell. Mitochondria used to be free living prokaryotic cells and have their own cDNA and ribosomes. Millions of years ago mitochondria formed an endosymbiotic relationship with eukaryotic cells.

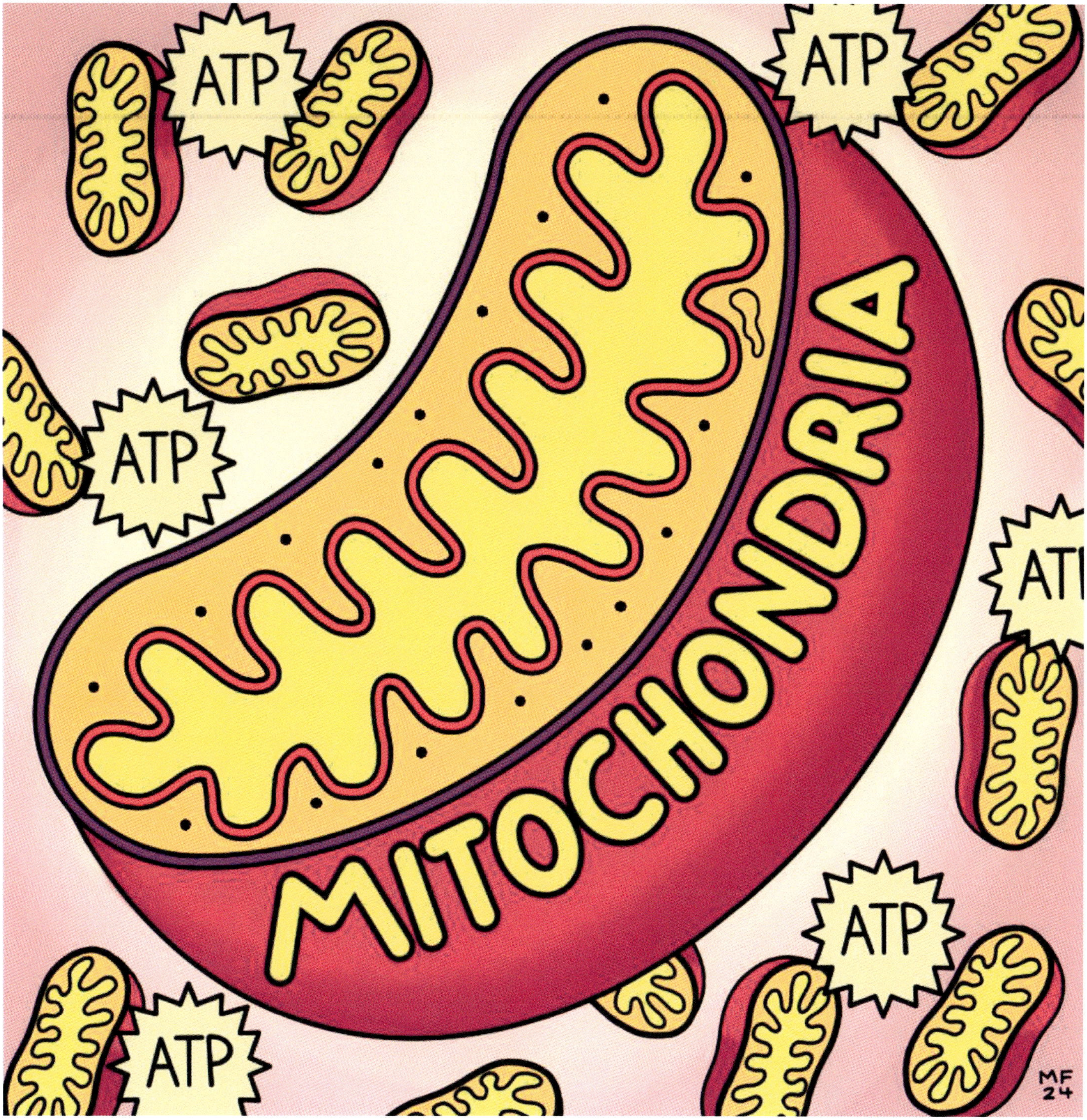
ATP
ATP
ATP
ATP
ATP
MITOCHONDRIA
MF
24

Chloroplasts: Membrane-bound organelles which are the site for photosynthesis found in eukaryotes. Chloroplasts are found in the cells of plants and plant-like protists. Photosynthesis is the process by which plants and other photosynthetic organisms convert light energy into chemical energy called glucose, while releasing oxygen into the atmosphere. Chloroplasts used to be free living prokaryotic cells, have their own cDNA, and ribosomes.

CHLOROPLAST
MAF24

Vesicle: Small temporary membrane-bound organelles found in eukaryotic cells which are used for transport, digestion, storage, and communication.

Vesicle
MF24

Centrioles: Cylindrical organelles found in animal cells, some fungus cells, and some protist cells. They are located near the nucleus, within a region called the centrosome. Centrioles play essential roles in organizing the microtubule cytoskeleton and are involved in various cellular processes, including cell division (mitosis and meiosis), and cell motility.

Centrioles
MF24

Lysosome: Membrane-bound organelle found in eukaryotes such as animal and protist cells. Lysosomes contain enzymes which break down and recycle macromolecules, cellular waste, food, and pathogens.

Lysosome
digestive enzymes

Peroxisome: Membrane-bound organelles found in eukaryotic cells which breakdown lipids using enzymes such as catalase to break down hydrogen peroxide into water and oxygen. Peroxisomes also break down nucleotides.

crystalline center
PEROXISOME
MF24

Amyloplast: a membrane-bound organelle found in plant cells. Amyloplasts synthesize starch granules from glucose and store starch. These starch granules serve as a long-term energy storage source for the plant. Amyloplasts are primarily found in storage tissues such as roots, tubers, and seeds. Amyloplasts are involved in sensing gravity in plant cells.

AMYLOPLAST

Protist cells: Specific eukaryotic cells that are the basic structural and functional units of organisms classified in the Protist Kingdom. Protists can be single cellular or multicellular. Protist cells, like all eukaryotic cells, are characterized by having membrane-bound organelles and a distinct nucleus containing the cell's genetic material. Protist cells have endoplasmic reticulum, peroxisomes, ribosomes, mitochondria, cytoplasm, and Golgi apparatus. Some protists have a macronucleus and a micronucleus. Others have multiple nuclei. Aquatic protists have contractile vacuoles to regulate water flowing into the single cell, while some also have a red eye spot to help detect light. Protist cells have a large range of unique organelles depending on whether they are plant-like, animal-like, or fungus-like protists.

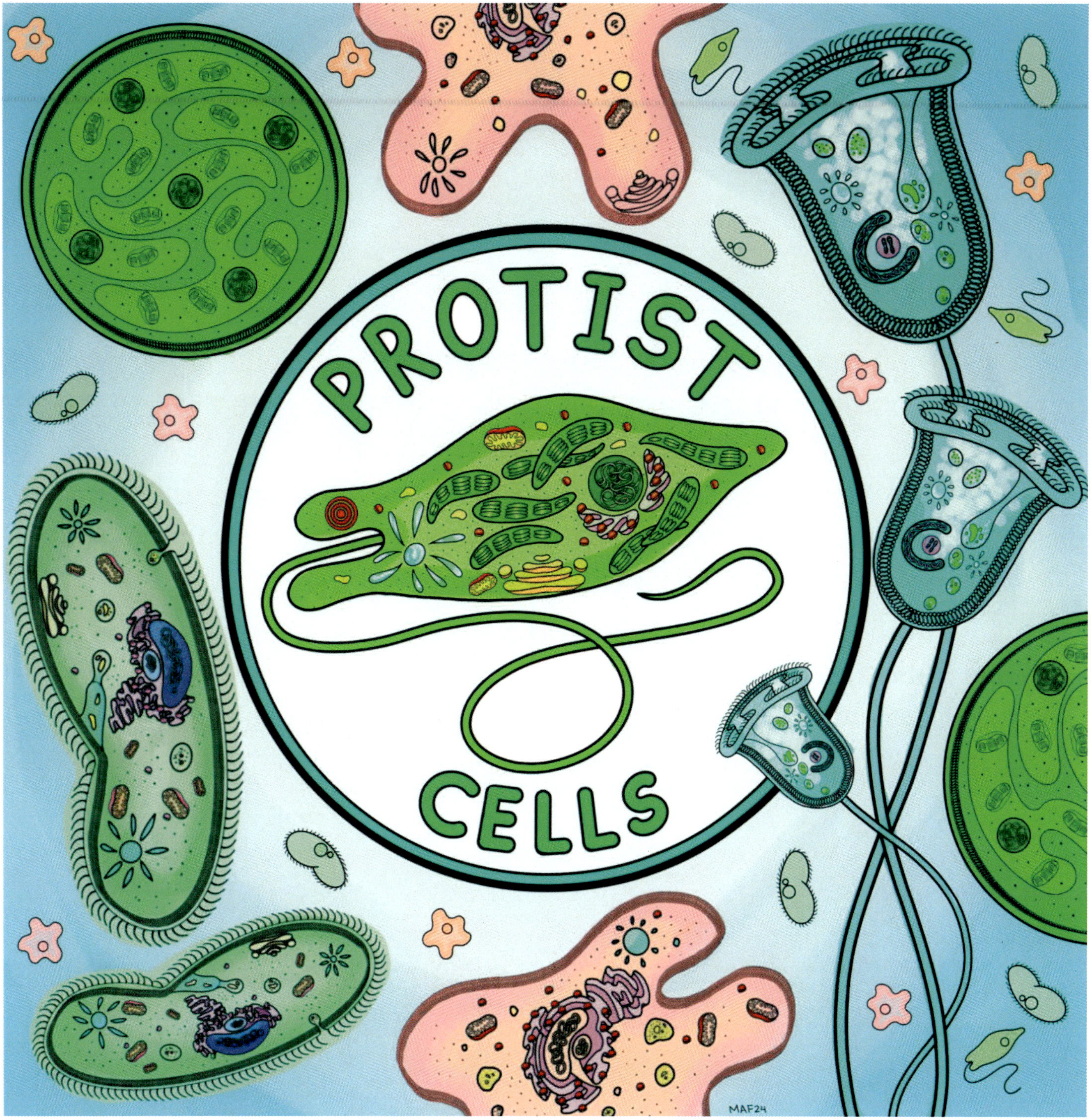
PROTIST
CELLS
MAF24

Plant cells: Specific eukaryotic cells that are the basic structural and functional units of organisms classified in the Plant Kingdom. Plant cells, like all eukaryotic cells, are characterized by having membrane-bound organelles and a distinct nucleus containing the cell's genetic material. Plant cells' most unique features are a cell wall made of cellulose, chloroplasts, and a central vacuole. Plant cells have endoplasmic reticulum, acidic vesicles, peroxisomes, ribosomes, mitochondria, cytoplasm, and Golgi apparatus. Specialized plant cells may have amyloplasts or chromoplasts.

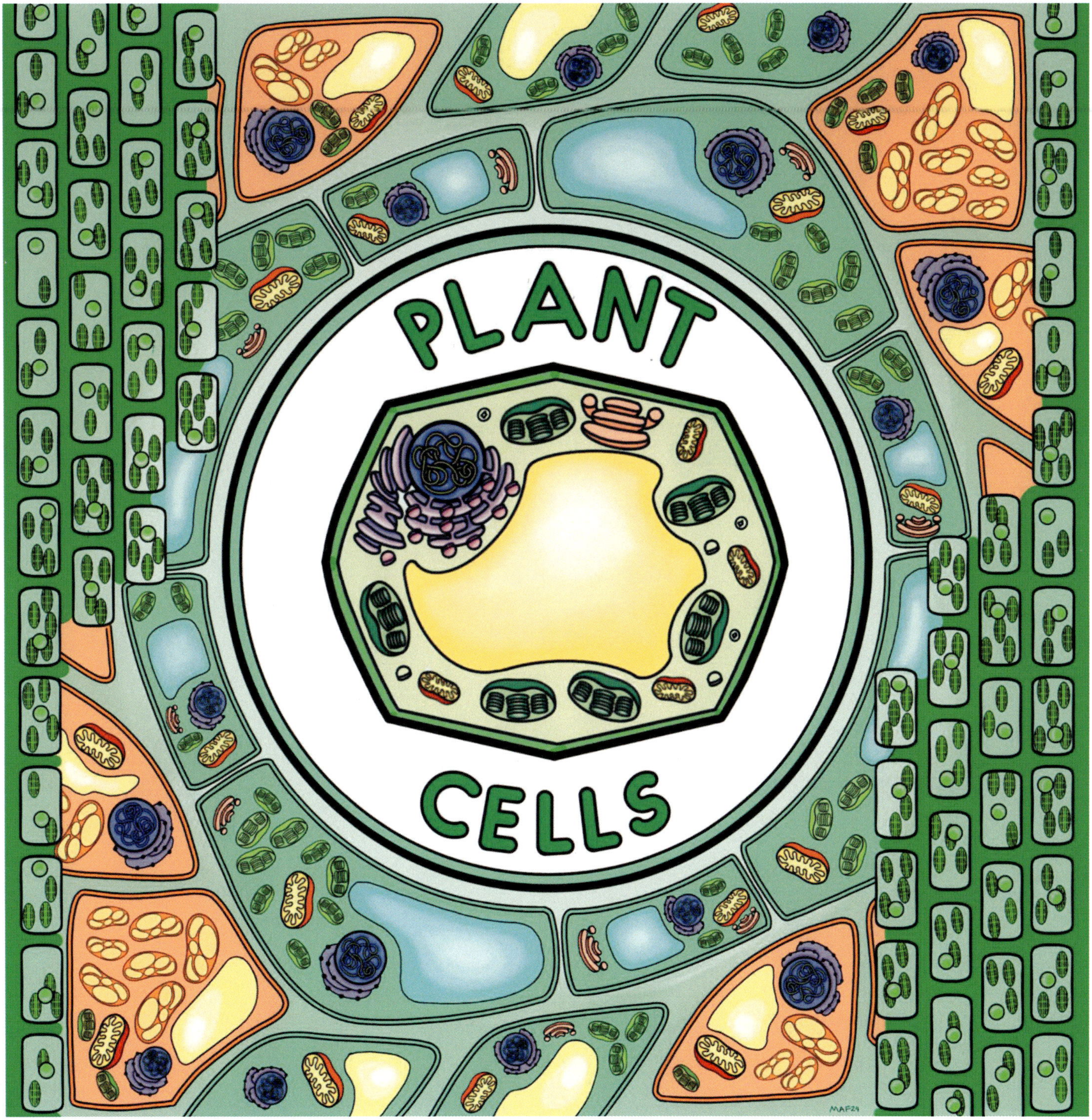
PLANT
CELLS
MAF24

Animal cells: Specific eukaryotic cells that are the basic structural and functional units of organisms classified in the Animal Kingdom. Animal cells, like all eukaryotic cells, are characterized by having membrane-bound organelles and a distinct nucleus containing the cell's genetic material. Animal cells have a cell membrane as their outer covering. There are no cell walls nor chloroplasts. Animal cells have centrioles, endoplasmic reticulum, lysosomes, peroxisomes, ribosomes, mitochondria, cytoplasm, and Golgi apparatus.

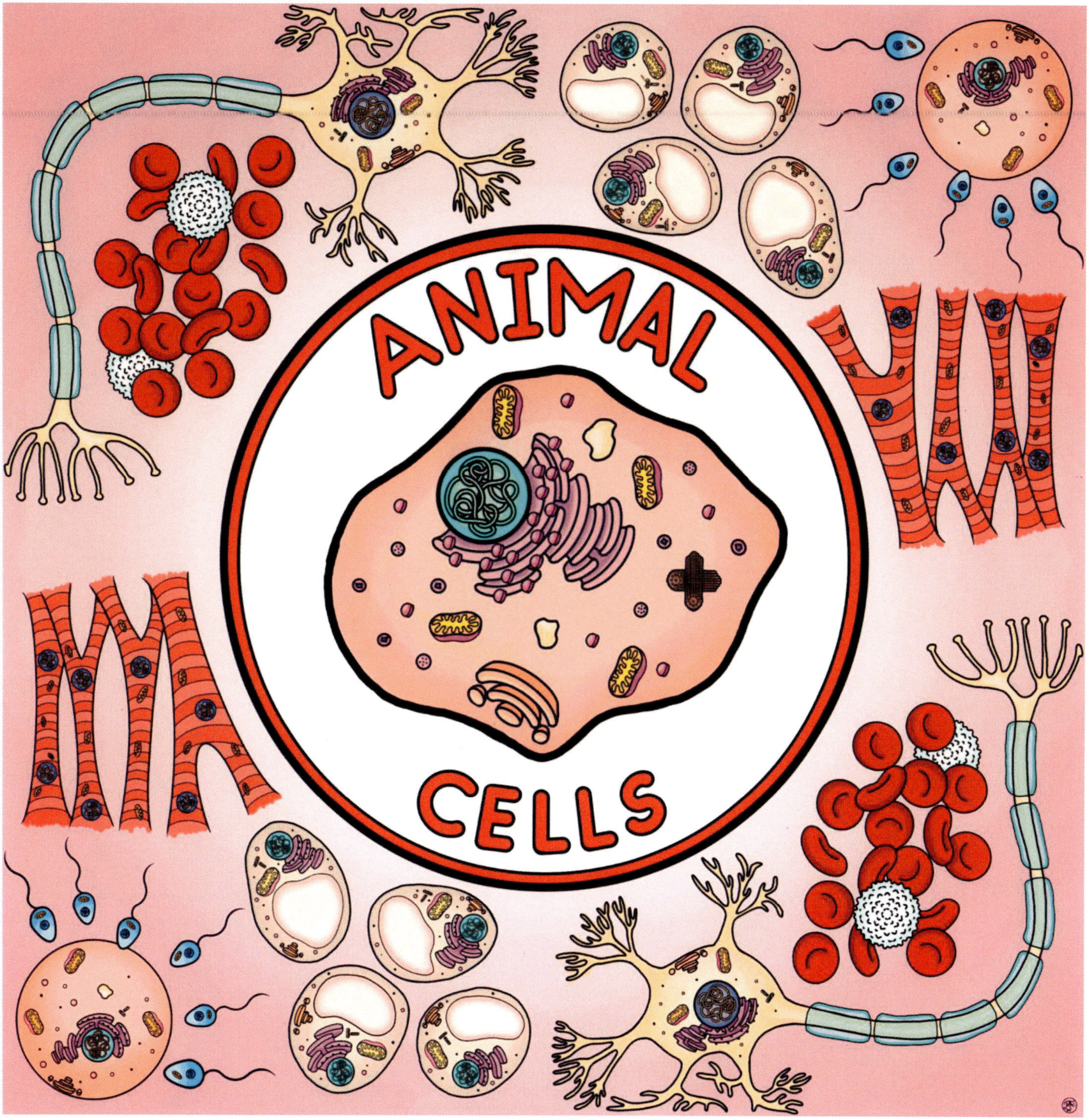
ANIMAL
CELLS

Fungus cell: A specific eukaryotic cell that is the basic structural and functional unit of organisms classified in the Fungus Kingdom. Fungus cells, like all eukaryotic cells, are characterized by having membrane-bound organelles and a distinct nucleus containing the cell's genetic material. Fungus cells have cell walls made of chitin as their outermost covering. There are no chloroplasts in fungus cells because they are heterotrophs. Fungus cells have endoplasmic reticulum, acidic vesicles, peroxisomes, ribosomes, mitochondria, cytoplasm, and Golgi apparatus.

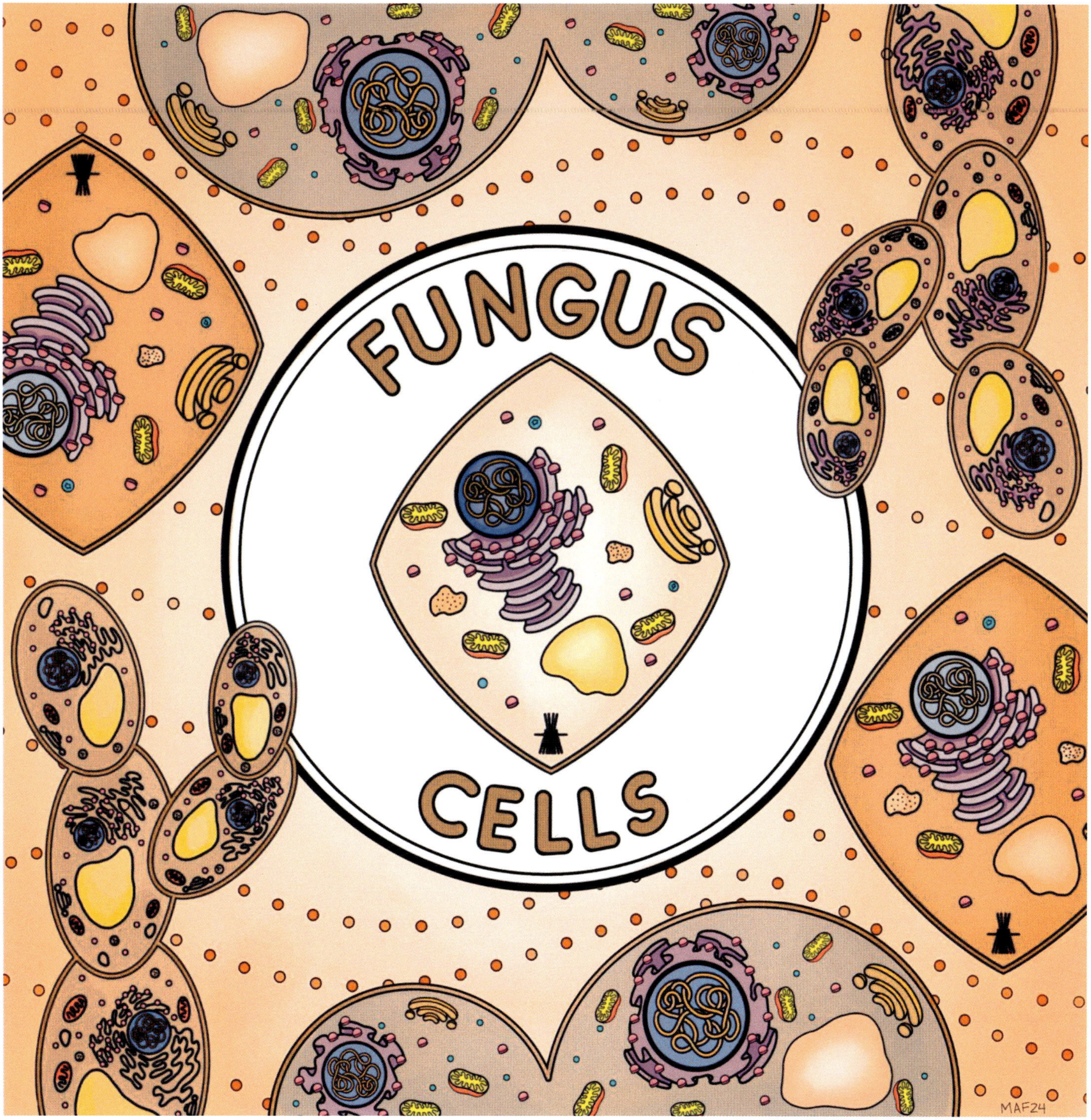
FUNGUS
CELLS
MAF24

Cell cycle: The series of events that take place in a cell leading to division and duplication. It consists of a sequence of phases during which the cell grows, replicates its DNA, and divides to produce two genetically identical daughter cells. The cell cycle is crucial for growth, development, tissue repair, and reproduction in all living organisms.

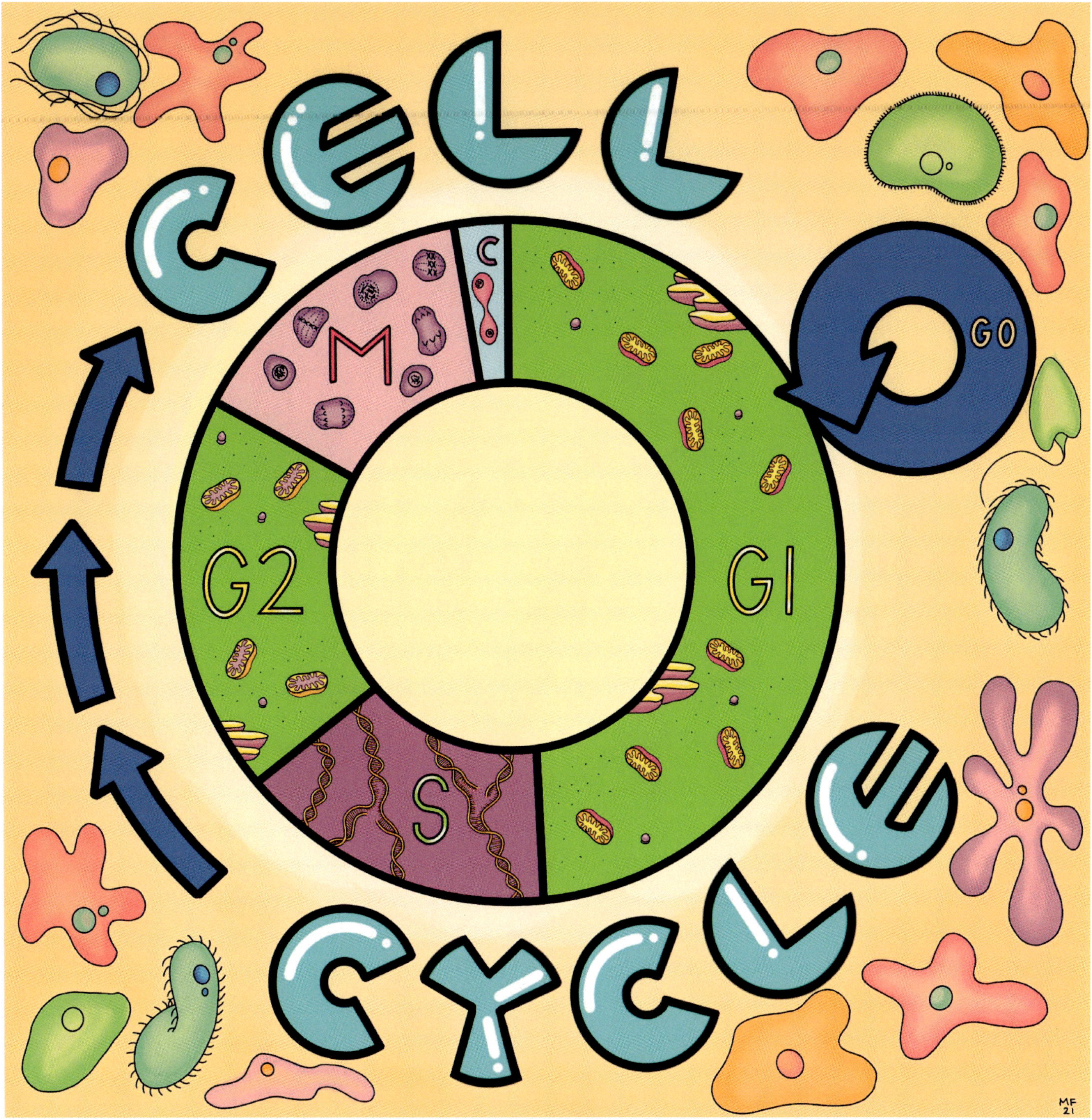
CELL
C
M
G0
G2
G1
S
CYCLE
MF
21

Interphase: The longest phase of the cell cycle, where the cell spends most of its life cycle. During interphase, a cell grows, develops, reproduces its DNA, and prepares for cell division. Interphase has three stages: Gap 1 (G1), Synthesis (S), and Gap 2 (G2). G1 is where the cell grows and develops. S is where the cell duplicated its genome. G2 is where the cell prepares for division if it is not in G0. G0 is where cells remain if they are no longer dividing.

Interphase
G1
S
G2
MF
21

Prophase: The first stage of mitosis and meiosis where the nuclear envelope dissolved while the chromatin condenses into chromosomes.

Prophase
MF
21

Metaphase: A stage within mitosis and meiosis where the chromosomes line up in the center of the cell before migrating to opposite poles of the cell.

METAPHASE
MF
21

Anaphase: A stage of cell division in both mitosis and meiosis. Mitosis has one phase of anaphase, while meiosis has two: anaphase I and anaphase II. Anaphase follows the metaphase stage and precedes telophase. During anaphase, the sister chromatids (identical copies of DNA) separate and move towards opposite poles of the cell.

ANAPHASE
MF
24

Telophase: The last stage of mitosis which precedes cytokinesis. During telophase, the chromosomes who have migrated into two groups at opposite poles of the cell will begin to decondense while new nuclear envelopes form around the genetic material. At the end there will be two new nuclei which will be divided into two cells by cytokinesis.

TELOPHASE
TELOPHASE
MF21

Cytokinesis: The final stage following the division of the nucleus in eukaryotic cells. During cytokinesis the parent cell divides into two new daughter cells. In mitosis there is one pass of cytokinesis resulting in two daughter cells. In meiosis there are two passes of cytokinesis, the first follows telophase I, the second follows telophase II resulting in four daughter cells.

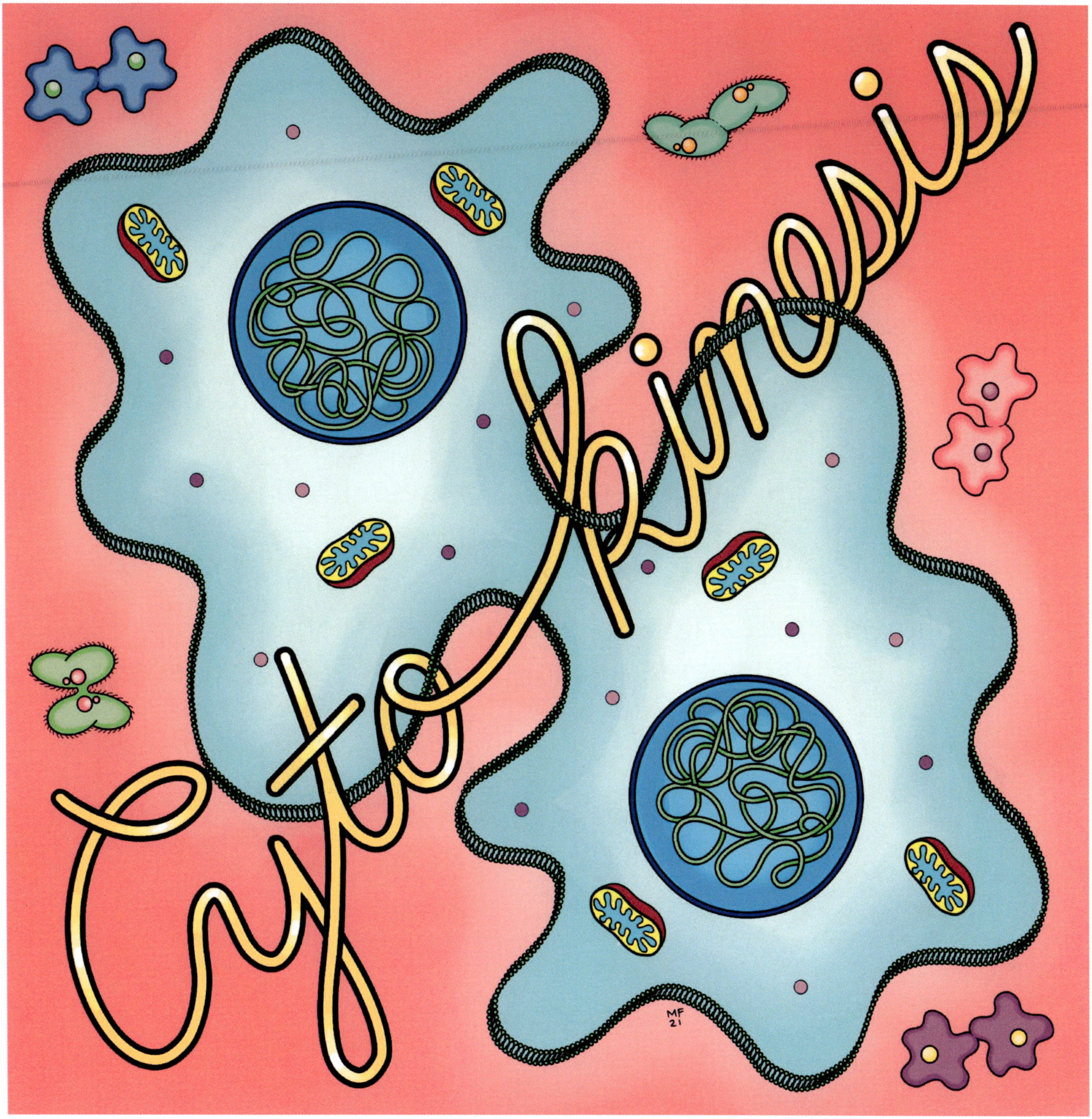
Cytokinesis
MF 21

Diploid: “Di” means two. “ploid” means a set of chromosomes. Diploid is used to describe eukaryotic cells that have two sets of chromosomes in their nucleus. One set is inherited from each parent. Mitosis creates diploid cells that are genetically identical and sometimes referred to as daughter cells.

DIPLOID
MF21

Cell checkpoints: Control mechanisms within the cell cycle that ensure proper progression through the stages of cell division. Checkpoints serve as regulatory points where the cell cycle can be paused or halted to allow for DNA repair, accurate chromosome segregation, and prevent the propagation of damaged or abnormal cells.

Cell
Checkpoints
M
C
G0
G2
G1
S
MF
21

Benign Tumor: A mass of cells that grows in a localized area of the body but does not invade nearby tissues or spread to other parts of the body. Benign tumors typically do not pose a significant threat to health, although they can cause symptoms depending on their size and location. Benign tumors can normally be safely removed.

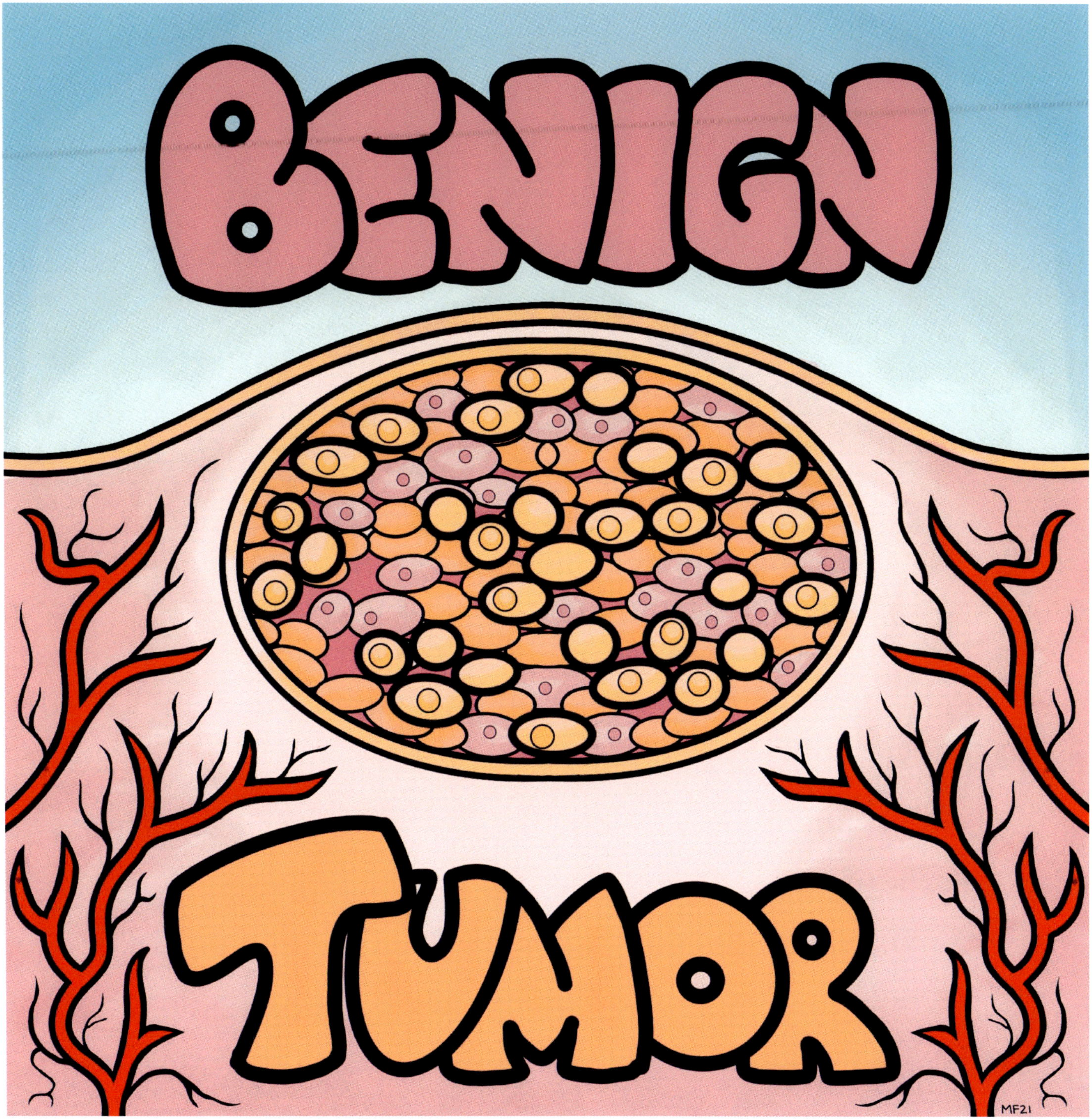
BENIGN
TUMOR
MF21

Malignant tumor: Abnormal mass of cells which divide and grow too quickly due to mutations in the DNA. Malignant tumors metastasize, which means they spread into surrounding tissues and can spread into new regions of the organism's body.

MALIGNANT
TUMOR
MF
21

Made in the USA
Coppell, TX
05 November 2026

11315625R00048